Praise

"Geld is gold. Provocative in a good way... the author 'goes there,' with intellect and without reluctance. I bet she enjoyed writing it as much as I enjoyed reading it."

—Jeff Jaya Sims, LMT

"Breathtaking, alive, heartfelt. A real game changer."

—David Matthew Brown, The Lion

"A beautiful book."

"Our bodies are not mostly 'out there.' They're 'in here.' Read this important book, trust yourself and listen to your body. The worst thing that can happen is that you fall in love with it. Oh, what a world that would be!"

—Anne-Marie Duchêne, Art of Alignment

"Inspiring treatise on wise and passionate inhabitation of one's own body! This is a lovely piece of writing combining the art and science, even spirit of inhabiting our body with passion and intention. In an engaging and whimsical adventure different ideas about experiencing pleasure and the joy of life are unspooled in a manner useful to anyone. Beautifully put and thoughtful, like the old adage "charity begins at home" this text puts us in our own bodies first, responsible for keeping channels open, energy flowing, and life renewing."

—Randy Polumbo, PLANT

"Protective armor has its place, particularly in hazardous environments. However its greatest value is recognizing when you no longer need it. Natalie Geld navigates this paradigm like a laser."

–Jeremiah Sullivan, SharkArmor Tech

"Excellent, an extraordinary way of seeing and knowing our bodies. Geld offers a... different vision of who we are as human beings."

—Frances G. Aponte Carrero, Artist

"Both an interactive experience and an insightful book, [Sensual Intelligence] is an amazing work of science and art. This book is an amazing look inside yourself, and also an interactive experience, where you're experimenting and experiencing your own body in new ways as you read. It's insightful and fun, this is a must for anyone who wants to try new sensual and sensory experiences, and I pity those who don't."

—Dylan G. Bank, Merlion Entertainment

"This is an enjoyable romp with information, perspective, humor, healthy openness, and optimism. I have been studying and practicing presence based coaching and this area of creating new habits and the physical rewiring that takes place is of particular interest. Recommended for those interested in a much-overlooked area of human development."

—Amazon Review

"A real turn on!"

"I loved reading Sensual Intelligence. It was engaging, conversational, and relatable. I felt like the author was talking with me. Her premise was well laid out. As the reading progressed and I followed the suggestions, I was enraptured, amused, energized, and really turned on. Afterward, I felt as if I was hanging in the afterglow conversation with my Lover -- Me."

—Amazon Review

"Promises a greater level of self understanding and Love for the human experience... a profound gift. The gift is that you don't even realize that you're constricting, you don't even realize you're cutting off your own flow of creativity, vitality, happiness, and joy. This book provides ways I can take better care of myself, to be my own best friend, lover, and co-creator. I feel more mindful now in my body, how it's all connected, and in how I feel throughout the day. I highly recommend this book!"

—Amazon Review

"The unique insight on every page gave me a deeper understanding of how my body works. I've always felt in tune, but now I have a detailed perspective and notice things about myself that I wasn't' even aware of. I love the way the author takes you on a personal journey beneath your skin. It was fun to read. I loved her writing style and can't wait for her next book!!"

—Amazon Review

SENSUAL INTELLIGENCE

An introduction to your body's language

NATALIE GELD

CO-FOUNDER OF THE SOCIETY FOR MIND BRAIN SCIENCES

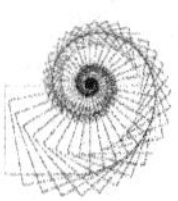

NAUTILUS PRESS ~ NEW YORK

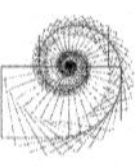

NAUTILUS PRESS ~ NEW YORK

2014 Nautilus Press Print Edition

SENSUAL INTELLIGENCE

Excerpt of Lifting the Skirt of Consciousness

Published in the United States by Nautilus Press.

WhyAreWeWhispering.com

This eBook contains an excerpt from the forthcoming book *Lifting the Skirt of Consciousness* by Natalie Geld. This excerpt has been set for this edition only and may not reflect the final content of the forthcoming edition.

This book is not intended to take the place of medical advice from a trained medical professional. Readers are advised to consult a physician or other qualified health professional regarding treatment of their medical problems. Neither the publisher nor the author takes any responsibility for any possible consequences for any treatment, action, or application of medicine, herb, or preparation to any person reading or following the information in this book.

Book Design by Shift Communication
Author Photo: Cynthia Smalley Photography
Audio Engineering by Joe Johnson at Red Wally Music

Library of Congress Cataloging-in Publication Data
Geld, Natalie 1964 –

Sensual Intelligence / Natalie Geld
1. Psychology – Emotions
2. Body, Mind & Spirit – Healing—General
3. Science – Life Sciences – Human Anatomy & Physiology
4. Self Empowerment —General

ISBN 978-0-9906218-2-9 (cloth)
ISBN 978-0-9906218-1-2 (paperback)

1 2 3 4 5 6 7 8 9 0

First Edition
NatalieGeld.com

Table of Contents

We are gods with anuses.

~Ernest Becker

1

SURFACING FOR AIR

HOW MANY OF US grip our sphincter, tense our abdomens, brace ourselves against the world as we walk around all day? Compelled to keep our bellies from spilling over our waistbands, tucking our tails under to maintain some sense of protection during this awkward stage between birth and death. Seriously, how many of us? Show of hands please?

My hand's up. I just caught myself doing it while I was typing. Perhaps you don't even realize your body is tensing up, armoring against what's going on in your world, and, more likely, your inner life. Repeatedly contracting these muscles restricts your breathing, blood flow, and your digestion. Limits the amount of oxygen, hormones, and neurochemicals that your entire body needs to function and flourish. Please wrap your fingers around your own throat and press in; until you're uncomfortable and have trouble breathing, it's like that.

Are you familiar with your sphincter? Contract it right now, as you read this. Breathe in as you slowly pull your sphincter muscle in and up into your body - like you're in an elevator and trying hard not to fart. Contract a bit higher... further still. Inhale more deeply as you pull up just a little more. Now release. It's likely you'll feel a tingling cascade of energy. Perhaps your eyes even brightened a little.

"Jerking ourselves off in the privacy of our own bodies!" Dr. Gil Hedley exclaimed as he directed all 54 of us in the Integral Anatomy Dissection Lab to slowly open our anuses, contract and relax them in unison, by gently raising each contraction higher and higher as you just did. We're pleasuring ourselves from deep inside, while we move our molecules. This conscious contraction and release stimulates and massages our muscles, fascia, fat, nerves, erectile tissue, veins, and

arteries in our pleasure rich pelvic region. In turn, stimulates our whole body; and it feels really good.

Beyond our own ass, we have over 50 different sphincter muscles throughout our body, some voluntary, like the one you contracted a moment ago, others on autopilot. There's the sphincter of Oddi, controlling digestive juices from the liver, pancreas, and gall bladder, into the duodenum. There are sphincters fairly unique to the mesentery surrounding your lovely guts. Your mesentery is made of two sheets of whisper thin, resilient tissue called peritoneum, sort of like a hammock, and connects parts of your small intestine to the back wall of the abdominal cavity. Touch your belly, and you touch your mesentery. Between each sheet of this supple fabric are blood vessels, lymph vessels, and nerves so your small intestine can move freely. Controlling blood flow are microscopic precapillary sphincters, cuffs of smooth muscle around each capillary acting as valves to regulate your blood flowing in and out of your stomach, intestines, uterus, aorta, even the artery to your right testicle.

There's a pupillary sphincter, which encircles the pupil of the iris — in your eye. You may have laid eyes on the sphincters in whales and dolphins that power their blowholes as they surface for air, releasing carbon dioxide, clearing and filling their lungs before diving down again. A sensational experience of awe and wonder - deeply resonating.

When you surface for air roughly 21,000 times a day, revel in the magnificence of your own breath. Inhale fully, and fill your lungs with wonder. Exhale audibly with a resonating… *awe.*

BATHING IN THE NATAL WATERS of our mother's womb our pleasure engine is already humming. Little boys have erections; girls suckle their thumbs, touch their faces, body, and breasts. Both sexes get pleasure from touching themselves during infancy without having to be taught. Everything goes in the mouth. Fingers reach for and wrap around anything possible.

Our brains get bigger as we evolve as a species, but our belief systems really need an overhaul. Little hands constantly get slapped away from their desire, accompanied with a threatening glare, or an emphatic "NO. Stop it. Don't touch that. Sit still." In the arms of our kin and communities, we're often conditioned that touching others and ourselves is wrong. It's not. Research has shown that the average preschooler gets thwarted in their desire about 20 times an hour. Neuropsychologist Dr. Rick Hanson concurs, "*Can you imagine being at work every 3 minutes… and someone stops you from doing what you want or gets in the way of you getting what you want? You'd get really pissed off after a while. That's like life as a kid."* It's how we're trained around touch, desire, play, and pleasure.

Scolding, shaming, emotional or physical neglect, and abuse reinforces this training; even indifference packs an emotional, self-critical wallop. Consequently, we become oppressed and limited in our ideas and abilities for self-pleasure. We become disconnected from our body's wisdom. And we suffer. Developing our sensual intelligence is key to unlearning this prohibitive conditioning.

If you think about it, we aren't taught how to communicate effectively. Loving touch is rarely demonstrated, and if so, not without some accompanying subtext or discredit. We're encouraged to 'suck it up' instead of learning to express how we really feel. We are, however, encouraged to eat more, lose weight, buy stuff we don't need, stuff ourselves with designer pills, suck down Marlboro Lights, Red Bull, play Angry Birds, and confess sins and sorrows to a bottle, bartender, and shrink. We numb ourselves externally and internally; contract ourselves woefully into our one-square-foot of "space." We live in our heads. We've become quite adept at our own dismemberment. Being out of touch and out of alignment feels like crap. Thing is, we don't even realize we need healing.

An abundance of undeniable evidence illustrates how stress affects our immune and nervous systems, causing painful symptoms, anxiety, depression, and disease.

Equally abundant, is the evidence that a few minutes of sensual human touch prompts a release of a number of "feel good" hormones, including serotonin, prolactin, and oxytocin. In this introduction, we'll take a hands-on tour of your body from head to toe and explore the art within. You will think about your body differently, even with mundane activities like sitting. *"You can decrease your own levels of the primary stress hormone cortisol, which blocks oxytocin's action in the female brain abruptly shutting off desire for physical touch,"* according to Dr. Louann Brizendine, the research scientist, clinician, and bestselling author who took on both genders in her spectacular books *The Female Brain* and *The Male Brain*. Another benefit? Cortisol regulates your appetite and carbohydrate cravings too, so your waistline will be pleased.

Knowledge has organizing power inherent in it.

DID YOU KNOW THE DISCONNECT WE HAVE with our bodies inhibits spiritual development? Perhaps you 'get' the idea of self-love intellectually, but don't 'feel' love. Ironically, we crave intimacy and fully armor against it.

Our body absorbs experience like a sponge. Even our lungs look and feel like plump, supple sponges. I know this because I spend hundreds of hours reverently exploring human anatomy. I marvel while breathing my own air into soft pink lungs of a human cadaver in the anatomy lab. When filled with air they appear to fly, like 'wings' of a great gentle Manta Ray gliding through the sea. Look at your skin. Notice its scars and wound etchings on your hands, your forearms, knuckles, and your thumb. If you peel off your clothes right now, how many healing marks could you count? Each one is a work of art, your cells reinvesting themselves to protect you. Each mark looks and feels different for the effort, different for the pull of skin over joints. Different for how it was cared for, or neglected. Imagine peeling away your artful skin, the largest organ through which we breathe. How many wound etchings can you imagine live inside? Each layer is intimately connected with the others.

If we could peel our skin, we would be surprised to reveal how profoundly interwoven we are -- every layer's interdependence stunningly clear. Intimacy begins with our sensual intelligence. Awareness and engagement, rather than fixation and attachment to our body, is power, deeply felt.

I've been wrestling with the disparity of communication in our medical and mental health fields my entire life. I've been honored to present my perspective, and interview some of our most lucid minds at the Science & Nonduality and Towards a Science of Consciousness conferences. I wanted to know why we feel so stressed and disconnected, even when practicing mindfulness. I wanted to explore why our sensual intelligence, our body, and our sexuality, is still left out of most conversations for raising consciousness and improving our experience of living. I wanted to know how we could evolve. What was missing? And you know they were glad I asked.

I interviewed Dr. Cassandra Vieten. She's extraordinary – a licensed clinical psychologist, President and CEO of the Institute of Noetic Sciences, and author of *Mindful Motherhood: Practical Tools for Staying Sane During Pregnancy and Your Child's First Year,* coauthor with Marilyn Schlitz of *Living Deeply: The Art and Science of Transformation in Everyday Life,* an avid soccer player, and a mom. Say that three times fast.

Genuine and direct, her energy and big blue eyes are quietly electric. Cassi blew into our interview room to discuss the truth about consciousness practices and our human experience. She embodied her work by being absolutely present, captivated me with her candor and said,

"What you are talking about is such a fresh topic, it's still so taboo. But for a few of the traditions of Tantra, there has been this idea that we have adopted now in the West that it's about catapulting ourselves out of mundane reality. It fits right into our whole Protestant, and Catholic, and patriarchal world-view that there is sin, our body and sexuality are dirty and mundane. There is just as much of

that in Eastern philosophies. They don't frame it the same way, but say it's a lesser world, and I really don't think it is.

I think that in this lifetime, in this incarnation that we are all sharing together, there is this unbelievable meeting of consciousness and matter. There is a meeting of energy and form, of immaterial and material, of subjective and objective. It's right at that meeting place where the action is. To fall too far on the material side is imbalanced. You are missing the joy, the richness, the beauty, the wonder, and curiosity of life. That's what life is all about. If you fall way over to the completely transcendent end of things (which I think is tempting because we have been so hyper-focused on the material that all of us in the consciousness world want to pull it all the way over there), the problem there is you forget about blood, and guts, and juice, and bodies, and sweat, and sex, and nursing. That's a mistake. We are not here to transcend this – it is how we are living right now. It's consciousness through form and form through consciousness. That doesn't mean that the way that I love my child, or the way I am attached to people I adore in this life, or the beauty that's created in a piece of art is any less than transcendent consciousness.

Mindfulness in motherhood for me was transformative; not only mindfulness transforming my motherhood, but motherhood transforming my mindfulness practice. There is wisdom in learning to be calm and neutral, to observe your experience in equanimity. But what I learned in the pregnancy process, the birthing process, the nursing process, and now even in the parenting process, it's about embodiment! It is all right here — there is milk and there is blood and there are tears and there is vomit — this is not excluded. In fact, this is what the journey of mindfulness is all about.

Mindfulness is about learning to approach all of your experiences, as much as possible with openness, curiosity, and compassion. Learning to ride the waves rather than resisting them and getting battered about in the process."

To sum it up, I like the way Cassi says, *"Consciousness practices offer richness, juiciness, depth, meaning, purpose, pain, awe, wonder, joy -- all of it. The whole enchilada."*

I talked with Kevin Krycka, PsyD, a Clinical Psychologist, Professor of Psychology, and Graduate Program Director at Seattle University. A warm and inviting human being, Kevin's levity is contagious. My face was as luminous as his while he spoke,

"If you are not a true human being leave this gathering, half-heartedness does not equal majesty. You know, that is an imperfect translation of Rumi. But what I take from it is that this whole business that we call spirituality, even something specific like enlightenment, is about being a human.

It is not leaving humanness. It is not being caught up in transitions, and transcendence, and the obliteration of ego. It's a support for us in this conversation of being human. What I see as depression, anxiety, or even pathology is a disposition away from one's unity consciousness.

Now, there are varying degrees of that. The smallest degree is called stress. Stress is experienced when we move slightly to the side, but only slightly, to the nature of who we are.

First and foremost, we are organic beings living in an organic environment. Everything about us speaks to our organicity. So, the further we become detached from our organism-ness, our flesh and blood-ness; detached from the reverie of tissue, and sinew, and brain patterns, and smiles, and laughter, and sexuality -- the further we become disso-

ciated from these, in degrees, is when we actually experience stress, anxiety, and then further along depression. Then, perhaps even further along something like suicidality, or panic.

It is our responsibility, our intentionality to live with who we are, to discover who we are, and then to follow truthfully and honestly from that discovery to the next moment, and to the next."

Embracing real, raw, visceral experiences is imperative to our spiritual development, well being, and to experience our unity consciousness. Let's dive more deeply into receptivity and our integral nature, into this experience of our interconnectedness to all things.

I spoke with Dr. Zoran Josipovic, a fascinating neuroscientist at NYU. As the founder of the Nonduality Institute, his perspective is expansive and finely tuned. We've become good friends and colleagues while founding the Society for Mind Brain Sciences. We talked about the misconceptions regarding brain, body, mind, and spirit. How they are often inaccurately compartmentalized, held as separate entities.

Even in consciousness practices, our body is usually left out of the experience of realization, which limits spiritual growth. I asked him to explain why it is so important to understand our profound interconnectivity, and the immense joy we can miss by merely living in our heads.

"Of course, this is a complex and subtle issue not so easily reduced down, as everyone has some experience of pleasure in their body, no matter how disconnected they seem. The issue is the way in which we objectify our body in our experience." Zoran said, and continued,

"Brain is not something that exists separately, as if on it's own – we are a profoundly interconnected and complex system of biological processes in which the brain, body and environment continually interact.

In terms of contemplative practice, if one does not include the body, and one's practice is only to quiet down thoughts or emotions, or to 'transcend' the bodily experience, then one's realization will be incomplete, mainly in one's head.

The more your whole body is open, the deeper your spiritual realization will be, and the more enjoyment you'll have being inside your body. To be slightly metaphysical, it

has been said that one of the main qualities of consciousness itself is bliss, so you are opening to that bliss of consciousness throughout your whole body."

THE EMOTIONAL STORIES of your life are written on your body's cells. Whether traumatic and painful, or joy filled and ecstatic, unconscious emotions shape your experience. If you wonder why you keep making the same mistakes, why you feel numb, or emotionally volatile when touched, ask your body. It's honest.

Your flesh and blood is your diary.

Apparently, Nobelist Dr. Eric Kandel agrees.

"Memories are stored not only in the brain, but in a psychosomatic network extending into the body, particularly in the ubiquitous receptors between nerves and bundles of cell bodies called ganglia, which are distributed not just in and near the spinal cord, but all the way out along pathways to internal organs and the very surface of our skin."

We are the authors of our experience. What our brain produces and provides is determined by the quality of the language we allow our body to express.

When we touch a part of our body, we touch our life experience.

When we touch someone, we are touched.

Everything is connected.

Next time you use your video camera, notice the limited area it's recording. Then look around you 360 degrees, as far as you can smell, hear, see, sense, touch, and feel. Your own body trumps that high def camera in your hand. Your cells record absolutely everything, even stuff you aren't aware of, from the most subtle and sublime, to the extreme. You gotta clear the database.

2

TUNING YOUR FEELING FORK

MAYBE YOUR BELLY and your butt are relaxed, yet you wall your shoulders around your chest, lungs, and heart. Are your shoulders high and stiff around your throat? Your neck like cement, teeth clenching while your hands clutch the sore parts?

We live in highly agitated states. Constantly preparing for perceived needs and threats, and frustrated about what we don't have. Spiraling in critical inner dialogue spurs us to contract, contract, contract, often unconsciously. If we have experienced prolonged abuse, neglect, or volatile, traumatic events, we become conditioned for anxiety and develop holding patterns in order to endure.

We're taught the false nobility of suffering for achievement. Pleasure and play get sequestered; everything is hard work, including relationships, career, family, and fitness.

Stuck in perpetual stress cycles - including chronic imbalance of adrenaline and cortisol - we increase inflammation and deaden much of our sensitivity. Clench your fist. Really hard, seriously. Now release the grip just a tiny bit, then clench again. Take a look at your muscles, and how the long, once-supple bones and tendons in your forearm are rigid now. Touch your veins that bulge like angry snakes. Notice how all the fibers in your body are connected. Fascia, nerves, capillaries, fat, muscle, tissue, tendon, bone – everything is compressed – restricting blood and oxygen flow, including chemical messengers that enliven the place you're touching!

This compression cycle is a big reason why you may feel desensitized and flat-out uninterested, even irritated when touched. Not only do you feel little or nothing, it just doesn't feel good, period.

Additionally, when someone's hand reaches to touch you, it conveys the possibility of having to fulfill yet another need. So you contract, mentally power off before it happens, thereby blocking neurochemical pleasure messengers in anticipation of depleting your system further. This leads to a place where touch is something that you have to endure, rather than enjoy. Where you believe that to touch, and be touched, has to lead to somebody else's satisfaction of some kind, with zero pleasure for you, and more energy exerted. Eventually resentment brews, plus exhaustion and pain set in. You lose trust in your body's ability to feel good. And one of those neurochemical pleasure messengers is oxytocin - our trust-building chemical.

So, what happens to our bodies when we suppress our dissatisfaction and frustration? A lot. Emotions are tangible assets. Avoiding our needs and emotions increases anger and anxiety and fuels our stress engine. *"Stress and emotions continually lodged at the level of the receptor of our body-wide system block nerve pathways (think gridlock) and interrupt the smooth flow of information chemicals, a physiological condition of being stuck - in sadness, fear, frustration, anger,"* says Candice

Pert, PhD, neuroscientist and author of Molecules of Emotion: The Science Behind Mind-Body Medicine.

Emotion = movement. The word 'emotion' is derived from the French word émouvoir, based on the Latin emovere, where e means 'out' and movere means 'move.' Our emotions don't just kick back inside that box you shoved them into, arms crossed, waiting for you to deal with them. What we suppress ultimately gets expressed in dysfunctional, painful and self-sabotaging ways.

Emotions translate as neuropeptides, small protein like molecules interdependent to their corresponding receptors on cell membranes, vibrating rhythmically in varying frequencies throughout our body-wide communication systems. By trapping our emotions deep in the basement of our awareness, we stop a natural biological flow of information and build armor in the fabric of our flesh and blood. We fuse our musculature into armor, intensely compressing our nerves, tissues, fluids, and blood flow, causing inflammation, pain, and recycling tired patterns. Tissues adhere - actually stick together where there should be fluid movement. Hold your hands palm side up, then lay one inside the other and gently glide around in all directions. This is the natural sliding relationship of many of the layers of our inner world. Now grab the other hand tightly while it tries to glide. How does that feel?

When we learn our body's language, when we recognize how all life is made of the same raw materials and patterns, we create a framework for a more loving, effective, and pleasurable response. Dr. Pert adds, *"You need to learn the alphabet before you can learn to read. Amino acids are the letters. Peptides, including polypeptides and proteins, are the words made from these letters. And they all come together to make up a language that composes and directs every cell, organ, and system in your body."*

How can we experience the intimacy we crave with someone else, while we are disassociated from our own body?

In contrast, we feel no subtext, giant agenda, or manipulation at play when it comes to our pets. We know, deep down, our pets just think our opposable thumbs are wonderful - to open that can of food or scratch beyond their reach. So when we're near them we touch them, this stirs our love, and immediately stimulates our oxytocin flow. Endorphins flood our system, alleviating painful symptoms. The more we pet and touch; the more the pleasure chemicals accumulate in us. Stress free drugs! We feel them, we radiate warmth; it's palpable. We feel good. They feel good. All is right in the world for those moments. We're not shut down. There is trust. We love.

We are love.

Close your fist again, this time when you unfurl your fingers halfway, imagine a small, brown sparrow is laying there in your palm, looking up at you. If I ask you to clench your fist again, will you? If you do you'll crush and kill the sweet little bird. Can you see a tiny, newborn puppy or kitten caught in your fist?

Imagine your own emotions crying out to be heard and cared for, your chemical information messengers begging to be released.

You are sensorial genius. Restorative, warm laying on of your own hands melts your armor; affectionate touch enlivens every cell.

Don't wait for love; *create*.

You stimulate generous love hormones that alleviate anxiety, relax digestion and heart rate, and generate warmth and intimacy.

3

WHAT IS YOUR HOLDING PATTERN?

PERHAPS YOU HAVEN'T NOTICED before. Many of us clench our jaws, hold in our stomachs, tense our shoulders, arms, and hands, or lock our hips, legs, and knees all day long. Each of us has our own unique holding patterns: squeezing in the diaphragm and the solar plexus, compressing the rib cage and everything inside. Tucking in the lower abs and buttocks, tail between the legs. Collapsing at the center, curving your shoulders in and around your heart, blurring the lines between breast and belly. Which is yours? How can you consciously unfold and unfurl?

We all need thoughtful reminders. Imagine I'm tugging the invisible silver thread extending from your crown to your own unique star high above. Allow your body to rise and sway in this connection.

Have you ever noticed how you brace yourself so vigilantly? Stir awareness so that each time you reflexively curl inward, lock down, and contract, instead you can crack a smile, breathe in, and release this habit, unfurling upon each exhale. It may feel somewhat sore at first; keep at it, again and again. You won't break if you let go. You will experience more clarity, more energy. Food will taste better, and life will smell richer. You'll find your chest swell like a regal bird fluffing its feathers. You will also enjoy more receptivity, both to your inner wisdom and the generosity and kindness of those around you. Others will respond to you more positively, relaxed confidence is contagious.

Do you have a force field emanating from you? Deflecting admiration and discrediting compliments cuts generosity and kindness off before it touches you. Notice *your* pattern when someone reflects your beauty, admires your talent, contribution, or achievement – do you deflect it back and compliment them instead? Perhaps you negate the offer, and say something like, 'No, I'm not' which generally spurs the other person to up the stakes and try harder. It's possible you even degrade yourself aloud, crack a joke to destroy the loving gesture, along with this life affirming energy.

Dare to evolve and do nothing. Just be still. Quietly absorb these loving gifts, these beautiful reflections of you when you hear them. Breathe the offer in, and say a simple 'Thank you.' Allow yourself to be lovingly touched this way. Feel the blush rise in your cheeks. Give yourself permission to receive pleasure, to openly share in appreciation.

WHEN I DISCUSS SENSUAL INTELLIGENCE with scientists, they want me to call it somatic intelligence, based on the somatic sciences. A more credible choice perhaps, as the word sensual conveys sexual rapacity, and may undermine its importance. I understand somatic to be a safer word for scientists that need to abide societal, political, and institutional systems in place for centuries. They have a lot of responsibility; I respect their position. I also believe sensual, affectionate touch, and sexuality, to be a major driver of evolution. Last time I checked, evolution was pretty high on the importance list.

I hold no PhD (yet), but I do have a dictionary.

The term somatic has Greek roots, sōmatikos, and means 'of the body.' Merriam Webster also notes that somatic means 'relating to the walls of the body, not the inner organs.' Wikipedia says it is 'relating to the body distinct from the mind, soul, or spirit.' In medicine, somatic illness is bodily, not mental, illness. Mind and body are one, so how does that work?

Our body is sensorial genius. Sensual touch is affectionate, loving touch. But the word sensual has gotten a bad rap, and refers, often unfavorably, to sexual, carnal pursuits and preoccupations. (Why those are bad, I have no clue. Our sexuality is an integral, driving power of our beautiful human nature.) Sensual, according to my grand old Oxford Universal Dictionary first published in 1933, with its slate blue cover dangling from its pages, means 'lacking in moral restraints; lewd, or unchaste.' Oh, how Victorian. Taboo of our sexuality, control of its discourse, and control of our bodies are very powerful forces wielding immense distortion and havoc. Google offers 'materialistic; irreligious.' I'm just staring at my screen at that one.

'Fleshly; animal.' Good, now we're getting somewhere - we are human animals, after all. Most animals you come across preen, groom, and soothe themselves ritually. Watching them lick and rub their faces on their fur or feathers is hypnotic and calming. Here's another, 'arousing or exciting the senses or appetites; worldly.' Now we're talking. Even Oprah concurs, *"Passion is energy. Feel the power that comes from focusing on what excites you."* She's pretty worldly, moral, charitable, and powerful, wouldn't you say?

To be effective in your life, the quality of your communication matters, which begins with speaking your own body's language. We're in this skin until we die, so we really do need to learn our own ass from a hole in the ground.

VITAL ENERGY IS IMPERATIVE to our health and success, so let's qualify our belief about libido.

Anatomy books are helpful, but they leave a lot out. To vividly and accurately express our profound interconnectedness, I want to see, touch, experience, and feel beneath our skin, so I reverently dissect, layer by gorgeous layer, our beautiful human anatomy. Exploring inner space with Dr. Gil Hedley, artful founder of Somanautics Workshops, and the first *Atlas of Integral Anatomy,* is transcendent, pure poetry, and loads of fun. Gil always shares his unique genius with abandon. He reveals divine truth, while re-imagining the body. I'm honored to call him my dear friend. My fellow Somanauts are some of the most extraordinary healers, teachers, and specialists our world will ever see, and I will be introducing their approaches to you in this Sensual Intelligence series.

I spoke with Gil about the misinterpretation of libido and energy, and asked him why he felt our libido is actually our life force. His answer, *"Because it's the most basic energy. You know, it's charge... discharge."*

Right. People get hung up on it, referring only to sexual energy -- that libido is just centered in the genitalia.

Gil said, *"Yeah, exactly. See, it's really just life force. If you channel libido out through your crotch, then it's what we call libido.*

When you channel libido up through your crown, we call it spirituality. Channel libido through your mouth, and we call it teaching. Channel libido through your heart, we call it parenthood. It's all the same energy. We can direct it. We can. We don't realize this... we don't dare to think how many choices we have because we are so busy playing the victim. And a part of the victim's role is to believe we have no other choice, right?

Your heart is having orgasms in your chest daily but if that's too much pleasure for you to handle – you'll hold yourself against it."

When I move my molecules to open my own information channels, powerful energy spirals through my limbs, freeing from my fingertips. It's so cool, and whatever emotion arises, it moves right through. In "Maleficent," the filmmakers portrayed the feeling of this experience, its influence on environment, relationships, and well being, and Angelina Jolie embodies this energy movement beautifully.

Have you ever heard people gripe about how the law of attraction doesn't work for them? How affirmations help, but only so much, because they still sabotage their own happiness. That no matter how much positive thinking they do, they don't get what they want. Positive thinking isn't enough when we're dismembered from our bodily, felt experience. Our emotional memories get tucked away in the unconscious, influencing everything we think, feel, say, and do.

Our body communicates in subtle and profound ways; most often we are unaware of this language being expressed, it's effect on our relationships and experiences. We're out of alignment on so many levels, conditioned away from touch and natural, physical expression, away from our innate power and healing abilities.

It's risky business living honestly, letting chips fall where they may. But living in denial of our inner life and the surrounding environment we call our body is a prescription for exhaustion, malaise and depression, lousy relationships, even illness. If you believe that you can transform your life with the power of intention, and get the things you want using the law of attraction, why leave your sensual intelligence out of the game? The majority of our communication is with our body – sexuality, love, power... everything is connected.

SENSUAL INTELLIGENCE BEGINS with conscious awareness of how our physiological systems process emotions and information. We are walking chemistry sets, human science experiments, all. Our organic unified flow is a highly concerted effort that we are an indelible part of. Consciously moving our molecules restores vital energy, enlivens sensitivity to heal, to feel good, to feel love.

We trust our health and well being to distinctly separate specialists, most know a great deal about a specific part or organ, yet have little understanding of the human being they're treating, namely you. These specialized fields are growing in leaps and bounds with advances in technology, while our education systems continue to churn out licensed medical professionals steeped in outdated information; ill equipped to effectively treat our ills.

Most of Western Medicine still treats the mind and body as if they're separate, even with so much evidence for unity. *"One of the most fundamental misconceptions is that part of the brain is emotional and needs to be overcome by the rational, civilized parts of our brain, when in fact, thought and emotion are inseparable."* says Dr. Jay Giedd, Chief, Brain Imaging Section in the Child Psychiatry Branch at the National Institute for Mental Health.

As laser focused Dr. Dean Radin, Chief Research Scientist at IONS and author of *Entangled Minds* puts it, *"We tend to make separations about practically everything – different names for different objects. We also make separations of mind and body, but there really is no separation. The separation is in terms of the words that we use to describe things.*

Most people think about mind being somewhere in there, in our heads. There is an inkling of evidence suggesting that there is also mind in your stomach, because it's the second highest density of astrocytes in your body; and intelligence in your heart. There is intelligence everywhere in your body, everywhere. So where is the mind? A mind is not just in our heads, and it's not even just in the body. It's probably everywhere. So, is that big enough?"

Centuries will go by as science and medicine cling to old theories or methods, before agreeing to retire them. Evolution takes time. Thankfully, technology is beginning to demonstrate evidence of our profound integral nature. Contemplative Science is beginning to flourish, bridging East and West in progressive ways. Is the great Wild West catching up to the mystical East?

While these ideas are churning, I thought we could get a jump on things with the skin we're in, while we're still in it. Why wait for science and medicine to agree on a new curriculum when we can learn to speak our body's language now? Why squeeze and compress the life out of our flesh and blood before we have the opportunity to revel in it? To really know what we're made of.

We walk around in these legs,
take painstaking steps
to get under the skins of those closest to us,
yet spend minimal time,
if any,
getting under our very own.

LOOK AT YOUR HANDS, YOUR FINGERS

Now please place your book - open to this page - in your lap or on your desk.

Let your hands hang softly near your heart, elbows bent and resting against the sides of your ribs. Lightly and slowly skim your fingertips along the curves of your other hand. Glide one finger along the other, one digit at a time, from its base to your fingers tips. Feel along your knuckles' ridges and swells that frame the root of your finger bones. With steady and calm strokes, notice the skin between each knuckle, on each finger, how it seems to rise to your touch. It is more sensitive perhaps along the sides of your fingers and the soft connecting web, as you dip down between each. Offer each fingertip and thumb opportunity to glide, and notice if the sensation feels different. If you don't feel sensitivity, it's okay; touch anyway. There are no qualifications on this.

If stroking causes you tension, discomfort or a bit of fear - that's why we're here. Rest assured you can take these steps slowly, at your own pace. Please do continue, though. Whatever is happening - you're at 100% because you're doing it.

Trace along your blue veins, all the way up your wrist. They could be pale blue and smooth beneath your skin or thick as cables, rising above and creating valleys between them. See how they meander and branch to wrap along the underside? And how your flesh changes texture, color, and porosity? If your thumb grazes and follows the same path as your fingers, it may feel foreign, a little rougher across the paler skin.

Imagine the first time you held an infant, perhaps your own newborn child, and marveled – enraptured by the miracle of their supple little hands. Remember how they felt when those graceful fingers wrapped around yours?

Experience your flesh this way now, as you press lightly into a springy tendon or graceful bone. Dive your fingers into the pool of your palm, drawing circles around the lifelines or carving initials into the trenches etched there. How are you breathing right now? Perhaps your shoulders are high and stiff, or you're clenching your ribs or belly -- if you feel your energies getting locked up, play with your breath. Inhale deeply into your shoulders, chest, and abdomen so they expand and soften while your fingers continue to explore; then exhale with any sound your body wants to elicit. Feel your voice resonate inside your chest walls.

Through this kind of focused massage, both externally – by touching your skin and body – and internally – by expanding, contracting and massaging your inner world with your breath - you can release tension, awaken sensitivity and replenish energy reserves. You can touch and massage your whole body to great advantage. *"We can induce sedation and pain relief by touching and stroking any part,"* says Swedish medical doctor and physiologist Kerstin Uvnäs-Moberg. And we also can stimulate the neurochemicals of love.

Slide your fingertips across your cheekbones, jawline, lips, cheeks – as I write, I'm doing this exercise along with you and it gives me the chills! And now I'm scratching my scalp vigorously and the chills raise my shoulders to my ears and race down my arms and down my spine until I squeeze my PC muscle and quiver like a kitty cat, smiling brightly from deep inside. There's a beautiful burst of energy traveling north and south. It's wonderful to actually spend time generating love in your body with your own hands.

Imagine a dog or cat's sweet and funny face – a fuzzy, warm and cozy feeling engulfs you. Pets light up when you enter a room, and come alive when you 'see' them, as if they didn't exist until you arrived. When you stroke their fur along their spine, ears, neck, and belly, you feel love and joy because you are 'making' love exist literally. When you touch them, or anyone, you too are touched. You influence potent neurochemicals, hormones, and energy flow beyond the visible walls of your body's largest sense organ, your skin.

When we see and feel ourselves this same way, we raise our sensual intelligence and our flesh and blood will respond in beautiful ways, transmitting boundless energy that lives within us.

Do you feel like a prisoner in your own body, avoiding emotional minefields? During World War II, The United States Playing Card Company cooperated with the U.S. Government in creating clandestine decks given to POWs; these cards could be moistened and peeled apart to reveal escape maps. Our mind cooperates with its own sensual intelligence agency – and we have in our arsenal a bounty of sensory motor escape maps.

Agency is the capacity of acting, or of exerting power. Our sensual intelligence involves our ability to read and interpret sensation arising from within our own body - a super power if you will. Scientifically speaking, this is a brain function called interoception, mapped in your insular cortex. Touch your scalp near your right ear, your insula is located deep within the folds of cortex inside your brain. Slide your finger in about a knuckle or two and you'll hit gold. You can develop empathy and your insula just as you build muscle, your athletic prowess, or business acumen. And it's a good idea to do so. If you want to experience more warmth and empathy from others in your life, begin by developing this in yourself.

Neuroanatomist Dr. Bud Craig at Barrow Neurological Institute in Phoenix, Arizona says, *"The state of your mind and the state of your body unite in the right frontal insula."* Our insular cortex maps varying states of our body; it's command central for our sense of well being. Anxiety, pain, depression, and chronic stress are felt experiences. So the more viscerally aware we are the more emotionally fit we become, and the more pleasure we allow ourselves, and others, to feel.

"It's interesting, people who meditate for example, who pay attention to the internal sensations in their body, it's probably also true for dancers, people who do yoga, anyone who is really deeply in touch with their body, they will build up a part of the brain called insula," affirms neuropsychologist Dr. Rick Hanson. *"The insula senses the internal state of the body as well as our gut feelings. When we build that up in ourselves, we are more able to sense it*

and to resonate with it in other people. You are tuning into the internal sensation of your body and that enables you to read other people better, not just your lover, but also customers, other people you are working with, your boss. You build circuits. You know, the classic line is neurons that fire together, wire together. So fleeting, immaterial, seemingly spooky mental activity literally forges neural structure. From a practical standpoint, if you are wanting to elicit more empathy from your partner or be more empathic yourself, one of the great places to start is to tune into your own body, pay attention to your breathing, pay attention to what your body feels like, very simple things."

Shake off apathy and approach your needs and issues with humor and humility; truly care for yourself. A simple way to humility is to employ your drunk monkey. I mean use the side of you that is ignored daily. We tend to be clumsy when we attempt to use our less dominant, under-developed side. If you're right handed, you're likely to be stronger, more physically developed in every way on your right side. This warrior side takes care of everything, leads to protect your more vulnerable, softer left side. In many cases, but not all, swap this if you're left handed.

From the moment your eyes open each morning, choose to do something different, leading with the under-used side of your body. Brush your teeth with the other hand, crack eggs, whisk anything, wield sharp knives to chop veggies (carefully), type on your tiny phone keyboard, put pen to paper, blow dry your hair, shave, wipe your ass – anything and everything. Do this often. Observe your inner dialogue. Is there a propensity to give up and use your dominant side instead, so you don't feel vulnerable or inept? How does this feel? What is this experience like in comparison to your usual routine? When we physically unlearn conditioned patterns of behavior, we activate curiosity and creativity.

Oh, and do show compassion and gratitude for the skin you're in - your kingdom of heaven - actually say thank you to your body for taking care of, and putting up with you all these years.

4

TOUCHING ON YOUR OWN RESISTANCE

WHAT DO WE DO if our own touch makes us feel ugly, stupid, or ashamed? I talked with Dr. Rick Hanson, founder of the Wellspring Institute for Neuroscience and Contemplative Wisdom, and author of *Buddha's Brain: The Practical Neuroscience of Happiness, Love, and Wisdom* about why our negative, fearful experiences arise so prominently and bully good feelings, and thoughts out of the way. Rick is a lightening shower of informed, positive energy, generously enlivening every moment. I asked Rick 'when negative thoughts and emotions are triggered during an experience, how can we neutralize a feeling of fear or discomfort so we can feel pleasure?'

"That is a great question Natalie. I think that's the question actually. Because, we are animals, right. We are at the top of the food chain but our animal ancestors have been at this for millions of years. Stone tool using fore-bearers of ours came along 2.5 million years ago. They had brains about a third our size but they were like us in the sense that they loved, they hated, they fought, they struggled, they dreamed, and they made stone tools. I can't make a stone tool, can you?

Over that long journey, it was so much more important to pay attention to sticks, and avoid sticks, than it was to get carrots. So the brain has developed what's called a negativity bias. In other words, it's like Velcro for negative experiences, but Teflon for positive ones; they just sort of slide right through.

Literally there are numbers of circuits in the brain that preferentially scan for, react to, store, and then recall negative experiences. So as a result, as we go through life most of what happens is positive or neutral, but what's negative sticks to us, where the rest of it just slips through.

The antidote is to deliberately look out there in the world so that positive experiences will stick to your ribs and as a result, they go into what's called implicit memory. It's not so much memory for specific events – it's memory for what things feel like.

Unfortunately, because of the negativity bias, the growing pile of negative experiences and implicit memories shades our experience in a negative direction – unless we deliberately antidote that by really paying attention to positive experiences – and what I call the 3 steps to taking in the good.

Step 1 is to turn positive events into positive experiences by really paying attention to positive experiences of your daily life. Feel them. Let them affect you.

Step 2 is to savor these positive events, to really relish them because it takes 15 to 30 seconds for the neurons that are firing together to wire together under ordinary circumstances. So you hold this positive event in your mind, you let it fill your body. You notice it -- that's the second step.

The 3rd step is to sense and intend that this positive experience is soaking into you; it's sticking to your ribs. It's kind of like golden syrup going down inside. And as a result, over time, you are going to start building up a posi-

tive sense of yourself, a more positive mood and greater confidence, with greater resilience for dealing with trauma or difficulty.

When you tilt toward what's positive, you're actually righting a neurological imbalance. And you're giving yourself today the caring and encouragement you should have received as a child, but perhaps didn't get in full measure."

5

RECIPROCITY

A FEW CONSCIOUS, dedicated self-pleasure moments - mindfully, with our thought and our fingers - are essential for well being because they unlock the physiological effects of stress. These moments inspire warmth, trust, and a calm relationship with ourselves. Look, when we bump our elbow hard enough to feel the shooting pain through our body, we immediately reach for the bruised place and rub it. Soothe the pain with deep breaths and tenderness. Sure, a flood of expletives will probably spill out, but hey, a well-timed 'Fuck!' does wonders. Point is we still touch ourselves when we feel pain. So, when painful experiences or uncomfortable thoughts and emotions arise, they're asking for your tenderness, your loving touch to soothe and heal the bruises, wherever they reside.

The upshot? We actually expand our capacity for love.

Just ask Anne-Marie Duchêne, the buoyantly powerful and intuitive Founder of Art of Alignment, uniquely empowering bodywork practitioner, and psychosomatic therapist, who says, *"As someone who touches bodies all day long, my own relationship with touch has greatly expanded over the past several years. At first, touching was almost uncomfortable. Yes, even for a so-called body worker. Interestingly enough, touching was not something that we did in my home as I was growing up. We hugged from afar, and smiled, and loved one another but no touching, please.*

People frequently visit me because they have a physical ailment and they want a solution, an intelligible explanation for what's going on with their bodies. Thing is, physical ailments do not occur without a psycho-emotive correlation. They never do. Everything is connected remember? The only trouble with recognizing one's psycho-

emotive relationship to their physical ailment is that it requires a little bit more of a dig: a sometimes mysterious and never before excavated archaeological dig. It can be intimidating and yes, it's work. But isn't anything valuable worth a little bit of work? Doing the work is the exciting part! That allows us to get to know ourselves better, and therefore know the other better. It supports our human evolution.

It is very important to give pain a voice. What's better still, and I believe most necessary, is a loving touch.

Why is a loving touch better? It immediately relaxes your nervous system. There is no fighting, no tightening of the under layers. Instead, there is full surrender, the type of surrender that leaves you connected and whole. The type of surrender where your tissues, muscles, and bones de-tensify, allowing you a private moment of deep awareness and appreciation."

YOU KNOW HOW IT FEELS uncomfortable when you first lay down for a massage? How you jump, or flinch a little bit, when a sore spot is touched? Then after a few minutes or so, you begin to relax and open to the experience – you're primed and begin to release the stored tension. It's like that when you begin to touch your body. It feels weird. Go ahead, laugh. Blush. Observe anger or rage. And please, by all means, do cry if tears well up. It's cleansing, and that feels good too.

Perhaps you're asking yourself, 'So, even if I don't have a mate, a lover in my life, I can actually be reverent with myself, and that's going to create an awakening?' *Yes. Absolutely, yes.*

Caroline Muir agrees, *"I am saying that. Before you go to sleep at night or when you wake up in the morning, touch yourself in a reverent way. That kind of reverence and that kind of art form doesn't have to have the goal of orgasm, it's simply a loving touch, something we all crave and need."*

Caroline is the best selling author of *Tantra Goddess, A Memoir of Sexual Awakening,* co-author of the best selling book, *The Art of Conscious Loving,* and a leading Tantra Yoga expert for 25 years. Fierce and elegant, she is transforming world-views from the inside out, and says, *"You may think 'Gosh, that could be lovely, but it's just not available.' We have to create our experience of love and awakening, we have to want such a thing."*

As you start to touch yourself, observe where your body is tender, sore, or numb. If an area triggers a painful memory and you feel yourself clenching against it – observe that too with empathy, a deep, slow breath and a softening stretch. Our pets are good role models for this. We're developing awareness, so if it feels overwhelming, rest, then shift your focus on cultivating the chemicals for healing.

Remember Einstein's theory of relativity, which demonstrates that the position of an observer will influence the phenomenon being observed, and will always affect the outcome of every observation. Your unified system that you live in is the most inspired phenomenon of all time.

Whether you have a vagina or a penis, explore the art and science of being human. The most challenging, and most rewarding relationship of all is with your self. Get to know you. Use your power to thaw your body that's been frozen and neglected all these years. You get to double down when you nurture yourself, because your love and kindness are never wasted, they are reinvested. They anoint the one who receives them (you), and anoint the giver (you).

Compressing our formidable energy stifles our capacity for love, intimacy, pleasure, joy, creativity, and our ability to express ourselves in healthy ways. It sucks. There's no magic pill to cure this, either. I can't simply anoint you with some ashes on your forehead and release you from angst to proclaim you an expert in joy and fulfillment.

What we are going to do – together – is unscrew the pressure valve and tune your consciousness to perceive your body from within.

Your task is not to seek love, but merely to seek and
find all the barriers within yourself
that you have built against it.

~Rumi

6

LISTENING TO YOUR BODY

GAZE AT YOUR fingertips. Sensory receptors called 'Merkel cells' discern the texture and structure of what they touch. These receptors are more closely packed on smaller fingers, in females generally. Which makes women better able to distinguish the shapes of the things they feel. We reach for and rub our backs, shoulders, and necks to relieve and soothe bruises and soreness, why do they get all the attention? Who pushed them in front of the line?

Let your fingertips take a tour, and see what happens when you touch different areas of your body. Be a scientist, archeologist, sleuth, or student.

Allow yourself to be influenced as you influence your flesh with your fingertips.

Does a certain pressure or region bring up tingly sensations? Emotions? Images of past events? A slight warming sensation? Indigestion? Arousal? No sensation at all? This is your baseline.

When do we take the time to lovingly touch ourselves just to touch? Finding time alone can be challenging, ways to approach these exercises include reading and absorbing them, then before you fall asleep or as you awaken in the morning, explore your body for a few moments. Sensually touching yourself during a shower or bath is natural, as your skin is already exposed, and warmed.

Engage yourself in this experience instead of channel-surfing crap TV or Googling yourself to sleep. Why not place your book in your lap, as you've just found time to read, and try it with me right now? Any way you choose is great. It isn't necessary to do all the steps that follow at once. Our connections are endless, and each time you touch you'll discover something new. Movement is meditation. Do one at a time if you like, promoting familiarity with an area of your body to awaken it. Play creatively; this is a starting point. And let the experience flow through you – no need to capture it and hold on, or judge it. What does your body want to say?

Just listen.

ALLOW YOUR TEETH TO PART COMPANY

Are you clenching? Attempting to hold your head up with your bottom teeth? Drop your jaw. Soften your tongue… it kind of floats there when you let go. Neither pressing into your palette, nor looking for something between your teeth, tracing every crevice obsessively until each tooth is outlined. When your jaw floats, so do you. Your shoulders naturally release their clench, your chest wall too.

Your heart and lungs are thanking you.

Your body's ability to maintain internal balance depends on a source of interoceptive maps involving receptors on your body's surface - your skin - including your teeth, gums, and tongue. Like clenching your fist allows nothing to pass through, clenching your teeth, compressing the life out of your tongue constantly ignites a cascade of closure throughout your whole body.

Mindfully softening and opening your jaw while lightly touching your lips together goes a long way baby.

YOUR SKIN

Your skin is your body's largest organ. It breathes. It emanates light and cleanses your body by releasing liquid and toxins intuitively. Embraces your body with just the right blend of thickness and elasticity for each region.

Profoundly interwoven, your skin doesn't peel away like a ripe banana as we might imagine; or see in movies or television. Feel the skin of a cantaloupe, for instance, and you are experiencing the underside of different areas of your skin, it's circular patterns so similar.

Deeply connected through fibrous layers, pierced by nerves and blood vessels, and interlaced with connective tissue, skin is our blanket, our beloved whole body hug, and our magic carpet to ride.

YOUR ARMS

Let your fingertips trace your blue veins meandering north and south along your other forearm. Play along these rivers and tributaries carrying your life's blood and oxygen inside them. Feel this energy. Press your thumbs and trigger fingers together, rub them vigorously and drag them up across your wrist, the center of your forearm, the inside of your bicep, and up across your shoulder. You're touching your median nerve, which controls your forearm and hand muscles, allowing your wrist, thumb, and fingers to bend and move.

Every time you grasp something, including this book or eReader you're holding, thank your median nerve. Feelings from these fingers, thumb, and fleshy palm at the base of these travel along this pathway too. When you're sore from typing and texting, bring your consciousness and warm touch here. This is just one of your many nerves, intricately intertwined with your bloodline.

Extend your arms like wings on either side. Rest a finger or two inside the cleft of your outstretched arm's elbow joint.

Paint circles there, where your humerus or upper arm bone, and your forearm's twin bones (radius and ulna) meet. Bend your arm slightly to feel these bones move. You have a glimmering superhighway of tendons and thin, specialized muscle tissue clover leafing beneath your skin.

Gently twist this forearm back and forth. As your forearm turns, so do these lovely bones. Feel them?

Now bend your arm again, this time allowing its fingertips to touch your shoulder and scratch – do your fingers resting inside your elbow feel the pressure?

Stretch your arm again and rake its entire surface with your fingernails. Notice subtle changes in sensitivity, even tug the funny, wrinkly elbow skin while you're at it.

YOUR CLAVICLE

Right above your breasts sit your collarbones. Like a swing set frame, we can wrap our fingers around them and gently hang there.

Wrap your fingers around your collarbone now, and stretch your neck up and away from either side, feeling your flesh tug beneath your fingers. Release your grip and slide and tap along your clavicle's stiff ridge.

Press lightly into the hollow cavern above, the smooth slope below. Remember to touch on both sides; they may feel different.

YOUR STERNUM

Explore the area between your breasts. This T-shaped long, flat breastbone protects your lovely heart, lungs, and life inside you. Tap in the center and on either side – the sound emanates differently.

Inhale and watch it rise as your lungs fill with air. Carefully press between your ribs. There are glistening, Saran wrap-thin fascial sheets webbing across them with thin layers of muscle between.

Slide your fingers directly center and feel the springy bone that seems to float. Explore and press very gently. This is your xyphoid process, and in Greek it means 'straight sword.' It resembles the tip of a sword.

Observed from above, it is beautiful as a butterfly.

YOUR BREASTS, YOUR CHEST

First Base. Fried eggs. Badoinkies. Airbags. Hooters. Jugs. Isaac Newtons. Ta Ta's. Bert and Ernie. Baby Buggies. Chesticles. Danny Devitos. Funbags. Samson and Goliath. Tits. The motherlode. Our currency. Our nurturing power. Our lovely, mismatched breasts glisten like sunshine beneath their fleshy surface.

Explore your breasts... we're all so primed to look for lumps and fueled with panic in case we find something odd that we avoid them. That's not our plan here. I want you to feel and explore and love the skin and flesh of your bosom. They're lovely. They're yours. Men, your chest is glorious so this means you too.

Rest the weight of them inside your palms as you hold and gently shake them side to side. Run your fingertips along the bumpy surface of your areola and your nipple. In the circle around nipples are Montgomery bodies or areolar skin glands. Their purpose is mysterious; perhaps they are there to lubricate your nipple and this area. Connect each dot with a feather light touch.

The male and female breasts have a similar nerve supply but the nerves lie closer together in men, and often are equally sensitive.

Nipples are tweaked or pinched a lot. Every day bound and suffocated by layers of clothing and contraptions to lift and separate. Tired and chaffed from the bounty of nursing. If your energy is flagging or you're feeling a little crummy, do this and warm energy will flow.

Slide your fingers along the edge up the surface of your nipple. Press gently down on its face. Swirl your fingertip ever so slightly there. Wrap your fingertips around your nipple and areola, and gently squeeze and release, squeeze and release without letting go, like a suckling infant. Gently twirl your fingertips around your now erect nipple. A wry smile on your face is worth the effort, along with a shower of oxytocin.

Slip your fingers and your whole palm around all of the flesh of your breast. Squeeze gently while tenderly caressing your fingertips in circular motions. Move them like an octopus all around, extending each finger as you do so. Rest your now warm hands on your breasts or chest, like sweet icing on moist cake.

Your heart and lungs are rhythmically massaging you as they fill with air and filter your life's blood. You can feel your heart now, and see its own arms extending into the wrap of your skin every time you see yourself.

Massage lightly under your arms with your thumbs, then down below your armpits. Many of our breasts spend a lot of time there anyway. When we lay down, we're thinking, woops, where'd they go?

Place the knuckle of your thumb under your breast, and gently lift. Then lay your other palm on your ribs below the fold of your breast where your underwire clings, sometimes viciously, to hold those babies up. Surgical scars especially need kneading to heal. Is it sore? Rub and slide your fingers across and in circles to soften and release adhesions.

Hold both breasts in your hands now – feel their warmth.

Your pectoral muscles above the fleshy part are interesting too. Hook your thumb into the same armpit and lay your fingers over the top and you'll feel this muscle tissue. Grab that muscle (between your breast and your shoulder) - and perhaps you can feel your ribs underneath.

YOUR BELLY

How do you feel about your belly? How do you like it just as it is - relaxed? How do you like it when you squeeze or hold it?

Do you love your belly or do you hate your belly? Everyone is going to have something to say, yay or nay.

Blanketing your innards – your stomach, liver, gall bladder, and intestines – is your very own grandma called your greater omentum. A thick, soft flap connecting at the very top, it is the guardian angel of healing for your tummy. Comprised of adipose (superficial fascia or fat), feathery fibrous connective tissue, and rich rivers and streams of warm blood and oxygen, she embraces your ailing organs. She moves intuitively to hug and hold and shelter you to help your organs heal. Hug her back by sliding your warm palms from one side of your belly to the other. Rub side to side, diagonally, and top to bottom; then in wide "mmm… yummy" circles from your tummy's periphery to your navel. You can press, hold, then release all over too.

While we're here, grab a handful of your belly fat and observe your thought process this triggers. Just listen. Now inhale as you roll your thumbs toward your fingers repeatedly. Allow a smile to rise as you slide your thumbs from your navel to your waist horizontally. You're caressing your fat, one of your more radiant organs of sensitivity and importance.

Your gastrointestinal tract is much like a bountiful garden you'd find below the sea – surprisingly beautiful, more like flowering coral instead of thin sausages. Areas of the small intestine include the duodenum, jejunum, and ileum, lovely words for our lovely guts. Your powerful sense of knowing - your gut instinct - is a voice that needs to be heard, not stifled by compression.

Your lovely guts also produce way more serotonin than your brain, and influence your immunity in a big way. Communication is a two way street between your belly and your brain via your central nervous system, so attentive TLC here goes a long way to improve your mood, relieve anxiety, help your brain – and your entire body - function optimally. The most potent pharmacy of all is you.

YOUR INNER THIGHS

Six muscles of varying sizes comprise your adductor muscle group. Your longus, brevi, and gracilis adductor muscles (along with their kin) span from your pubic bone along the inner part of the thighbone, or femur, to your knee. Intertwined with tendons and slick deep fascia, they look like broad egg noodles when lifted out of their bone cage.

Slide your fingers, and your entire palm along these inner thigh muscles and gently knead them with the heel of your palm from above - where they connect with your pelvic bone and labia. Move across the span of them, all the way to your knee to loosen, revive and warm them. Curve your fingers inward like a rake, and drag them up and down and in circular motions along this area.

Sync your breathing, slow and deep, to the rhythm of your touch.

YOUR HINEY

Just saying that out loud makes me smile. Say it: My Hiney. Our ass usually wields more frustration and insecurity than smiles; just buying a pair of jeans or walking to the bathroom from a restaurant dinner table can be nerve wracking.

Read this aloud: Posteriorly, the superior ramus of the ischium forms a large swelling, (I'll say) known as the tuberosity of the ischium (or ischial tuberosity), also known as the sitz bone, or as a pair, the sitting bones.

When we sit, our body weight is frequently placed upon the ischial tuberosity. The gluteus maximus (your butt muscle) covers it in the upright posture, but leaves it free in the seated position. That means whenever you're seated (long work hours at the desk, your car commute, church pews, during meals, on the hopper), everything in between your sitz bones and what you're sitting on is getting squished. The sitting bones are pressured between your body weight and the surface you place them on.

They're also storing all that self-loathing, sadness, anger, and shame. And we often clench our sphincter – our anus, our hiney hole – when we're nervous, fearful and self-protecting, anal or control-driven, struggling with embarrassment, IBS, or gastroenteric symptoms.

However, relaxing this area frequently is vital to healthy digestion, breathing, blood flow, and emotional balance, so reach behind and massage your own sitz bones – circling in deeper to where they get really tight and sore. Sit on your hands now. You can feel these bones just below your butt cheeks. Grind into your fingers. Melt your butt. A simple tool to use is a tennis or golf ball, or a small, firm (and clean!) dog ball. You can roll your hiney and sitz bones on them – this really works to open up stored, compressed energy.

YOUR HIPS, YOUR LEGS

These feminine swells set us apart from the more rapier lines of the male species. Swim your palms around your hips - press and knead there gently with your knuckles. A broad hand with a tacky contact will create drag through these layers. Your whole hand can be creating vectors of pull in a friendly way in the deeper tissue there.

Wrap your hands around them, press and grasp firmly. Inhale naturally; then exhale as you glide your fingertips and palms up and down them. This may bring up trickles or floods of discomfort – deep breaths now, strong and slow. This jointed area manages a lot of action with just your daily sitting, standing, walking, exercise, and bending; let alone from your hiding or lamenting them, engaging in sexual activity, or repetitive anal sphincter clenching.

Lay your fingertips on your hips. What do your Merkel cells think your hips have to say? Remember, pity is not empathy. Pity is also not love. Allow yourself to listen with empathy.

Sweep along your outer thighs. Swirl around and under your knees, so soft there.

Slide your fingers along the edge of your tibia, the larger long bone in your calf. It's tough. It could be the strongest weight bearing bone you have as it takes an axial force during walking about 5 times your bodyweight. Can you feel the flat hard edge down the center? Skin is thinner there. Now tap from top to bottom, hear that?

Rub your fingers vigorously all the way down it, across your ankle, your foot, and back to rest on your hips again.

THE GATEWAY TO YOUR PERSONAL POWER station, housing billions of concentrated nerve centers is ...

YOUR PUBIC BONE

Press your fingertips into the center of your fleshy, furry mons veneris - the triangle island of pubic hair - until you hit gold. This area can be sexually sensitive and protects your pubic bone from the impact of thrusting during sexual intercourse.

When you hit center, as you press in above the bone's ridge, you may meet someone there you really should get to know.

Hooked into your urogenital system is a pleasurable anchor point called the median umbilical ligament. There's another just like it, called medial, running like an inverted peace sign from your belly button to the place you're feeling right now. Could this be an embryological vestige – your yolk sack still lingering there? All I know is when the nuns told us touching our genitals was a sin; this was my get out of jail free pass.

As you massage, you may feel waves of pleasure. This is your pyramidalis muscle. Each one is unique in every way.

Feel the ledge of your pubic bone. Apply pressure slowly and inhale, hold for a moment, then release your fingers and your breath together. Nice and slow. Inch along your pubic bone and repeat this acupressure. Tug your pubic hairs gently.

Still pressing and breathing, follow the foundation of your pubic bone beneath your mound and inside the valley between your labia, or your penis, and your legs. Skin is very soft there.

Switch now between gliding, smoothing and pressing, and then extend along your inner thigh again. Sense these connections and textures. Mix it up. Tug your soft outer lips – your very regal labium majus, or supple skin at the base of your penis.

Gather up small bundles of this flesh between your thumbs, fore and middle fingers and roll tenderly, making the gesture for 'how much.' Walk your fingers along these areas.

Collapse them down and lay your entire palm on this area. Like an electric blanket, allow this warm energy to flow throughout your whole body.

YOUR PERINEUM

With your hands warmly blanketing your pubic hair (or soft skin if you are hair free), mons, your vulva or either side of your shaft - press your fingertips lightly into the flesh just below your vaginal opening spanning to your anus. It's behind your balls if you have a penis, (your boys or testicles if you will). Right there in the middle.

Referred to as Biffin's bridge, grundle, taint, or gooch (not to be confused with the Senator.) Also called your perineal body, it is known to some as the "anchor of the pelvis" and is one part of your body you don't want to miss. Smooth and soothe this field of dreams.

Feather your fingers out towards your inner thighs, up and down rhythmically, then circle and press in to feel it press back.

Quietly lay your hands, one atop the other, over your penis, over your vagina. Over these layers of spongy erectile tissue, muscle, fascia, blood, and nerve tangles feeding them. Warm and still with slow, luxurious breath now.

Breathe the energy into your hands, up through your supple wrist, forearm, biceps, shoulders, collarbone, and heart – golden light filling you through every pore. Radiant spiraling energy into your lovely guts and down your strong legs. Filling every plump organ, slithery nerve, mixed with your interstitial fluids lubricating and nourishing every cell of your body, now companion to your blood and bone.

Wiggle this light through the tips of your toes.

Go ahead and sweep down between your buttocks and say hey to your anal area, feel its pucker in the folds and fan out across your cheeks. Cradle and squeeze your ass with the warmth of your hands. Spank yourself if you're so inclined.

YOUR SPINE

Glide your fingers up your body to place them firmly at the base of your skull. Open your mouth and allow your head to fall back and rest on the ledge you created. Feel its weight – your head's pretty heavy. Let your throat relax and open, lots of energy trapped in there with all the things left unsaid.

Bring your head to its natural position and trace your fingers and thumbs as best you can to feel the length of your spine from top to bottom, all the way to your ass crack. Your supple spinal chord is your ancient information super highway, part of an older sensory tract developed over millions of years, transmitting crucial information to all areas of your brain and your body. Could back pain be related to restricting this flow?

To hold your liquid rich, soft and near translucent spinal chord in the palms of your hands is magical. I've done this several times. Feel yours now, and sway gently like soft reeds in a warm breeze.

YOUR CRANIUM

Now reach for your scalp and rake your fingernails from the nape of your neck upwards to the crown of your skull, then swirl them along the top. Start again, expanding your fingers all the way across your temples. Your feathery fans of muscles beneath your fingers are a sculptor's delight, thin and delicate, as they are strong.

Rest your fingers and thumbs along your brow.

Slowly glide them into your hairline (or where it used to be ;) – this is the frontal belly of epicranius muscle, or your frontalis muscle. Slide down behind your ears and glide your fingers to the back of your head, this is your temporalis muscle. Explore with varying pressure, tempo, direction, and patterns around your skull, behind your ears, gently tugging your earlobes, ears, and tufts of your hair.

Remember to open your mouth and relax your jaw to feel the full energizing, tingly effect generously distributing palpable energy. Drink it in.

Just as we began, slowly pull your sphincter muscle in and up, deep into your center. Melt into the flow… long slow breaths with each lift and release.

7

FEELING IS HEALING

BEING THANKFUL IS HUGE. Gratitude is powerful medicine. Appreciation magnetizes more to appreciate into your life. Give it a whirl and see. Share your admiration and gratitude by saying thank you aloud often so your appreciation resonates through every molecule, heals you and lights your cells from within. Your body will thank you back.

When your hands are occupied, use your mind's body to stimulate the inner feeling of an action. In other words, affirmation, visual imagery, prayer is all very helpful but only motor or kinesthetic imagery — mentally rehearsing an action again and again — will alter your body maps positively in the same way physical practice does. It's nearly identical to your brain's motor cortex, which we use during physical activity.

Athletes, musicians, dancers, yogis, surgeons, and healers use these body maps, practicing with kinesthetic imagery to great advantage, and to attain mastery. You'll feel much the same results by visually rehearsing this sensory exploration of your body. You'll develop your insula, relieve stress, and feel more supple. You will enjoy more empathy, love, self-awareness, and pleasure.

Remember Rick's advice, 15 – 30 seconds to marinate an experience. For lasting transformation, it also takes 10 to 14 days of visual or physical rehearsal to trigger the complex cascade of gene expression that allows your brain to change, and forge new neural pathways. Here's to engaging your sensual intelligence wherever you may be.

Thank you wholeheartedly for listening, and for sharing your valuable time with me. Until next time…

"Take your practiced powers and stretch them out until they span the chasm between two contradictions... For the god wants to know himself in you."

~Rainier Maria Rilke

EXCERPT FROM *LIFTING THE SKIRT OF CONSCIOUSNESS*

We're all here because somebody got laid. So why do so many of us have difficulty expressing our sexuality, enjoying sex without inhibition or fear, talking about our needs and desires with a lover?

Well, sex is hardly ever just about sex. Moreover, physicians who treat mind and body, scientists who study the brain, and consciousness transformation itself, somehow left sex out of their research. How can something like sexuality – our primary phenomenal experience that drives evolution – be left out of these conversations when our communication occurs through our body?

It can't.

Sexuality, love, power – everything is connected. Sexual energy contains our stories. Our body is the imprint of emotion. Their release liberates us to experience unexpected, unparalleled pleasure. Hiding our desires, sexual curiosities, innermost feelings, fears, and needs often causes that Holy Grail we call orgasm to become frustratingly elusive. *"Keeping sexuality out of our awareness is like trying to keep ping pong balls under water,"* as Taoist Master Mantak Chia says, because suppressing our natural biological flow of communication along with our powerful sexual impulses is a no-win, and exhausting, routine. Sexuality influences our thoughts, choices, and behavior; our belief systems and latent fears profoundly affect our emotional health and sexual expression.

Body and mind in silent battle.

There's more to satisfying sex than to stroke, lick, suck, or screw. Sex is more than mechanics; it's communication, and it's influenced by un-integrated emotions, experience, and conditioning. While candles, panties and porn are titillating props, real intimacy demands that we shift beyond merely the routines of sex to its full expression.

What is a healthy sexual expression?

How do we know if we need healing?

Co-founder of the Society for Mind Brain Sciences, author and educator Natalie Geld blends inspired, sensual storytelling, proven methods artfully expressed, and provocative conversations with thought leaders, scientists, doctors, clinicians, shamans, real men and women, and sexual pioneers.

In the dark about your sexuality? You're not alone.

Living from the waist up? Lift the skirt of consciousness to reveal the reverie of your flesh and blood.

Unravel this feedback loop of sexuality and stress so you can really feel love and have more fun in the skin you're in. Challenge your belief systems to unleash your inherent power. Open the floodgates of pleasure. Awaken receptivity and relieve stress to transform how you feel, live, and love.

Communicate effectively in your relationships – with your doctor, your lover, and your body!

Better health, better relationships, better sex?
Who wouldn't be motivated by the potential!

PRAISE FOR NATALIE GELD AND *LIFTING THE SKIRT OF CONSCIOUSNESS*

"A fearless exploration of human sexuality and consciousness. Ms. Geld's prose is intentionally brazen, literally flipping up the skirt of our still-somewhat Victorian sexual attitudes to reveal the complex physiological, psychological, behavioral and cultural factors that shape our sexuality. Prepare to be intellectually stimulated, sexually liberated and thoroughly entertained!"

–Peggy La Cerra, PhD, neuroscientist,
CSO of God's Speed, author of *The Origin of Minds*
(with Roger Bingham)

"Imagine the rigor of science, beauty of art, wisdom of contemplation, and sensuality of true intimacy rolled into one joy to read book. The wealth of the ideas and insights presented in this illuminating work will be of great benefit to all readers, encouraging a new way of living and being that has the power to transform our individual and collective lives."

–Shauna L. Shapiro, PhD,
Assoc. Professor of Counseling Psychology
at Santa Clara University,
co-author of *The Art and Science of Mindfulness*
(with Linda E. Carson, PhD)

"This book is the GPS for improved health and happiness. As a general internist and addiction specialist, I consider this book a must read for anyone seeking an epiphany on how to manage their good health. Doctors, like their patients, need Ms. Geld's landmark work to truly fulfill our Hippocratic Oath. Bravo Ms. Geld, where were you when I went to medical school?"

–David Kipper, MD,
author of *The Addiction Solution: Unraveling the Mysteries of Addiction through Cutting-Edge Brain Science*

"Natalie Geld's sharp intelligence and heartfelt enthusiasm support her deep commitment to spreading delight in a fully conscious manner. Her "communicable ease" regarding a fully integrated experience of human sexuality is contagious, and deserves our full attention."

–Gil Hedley, PhD,
founder of Somanautics Workshops, Inc.
and author of *The Atlas of Integral Anatomy*

"So where does sexuality fit in with our well being and spirituality? Conscious love is the ultimate expression of our nonduality. Read this magnificent book about true integration so you can evolve as conscious beings."

–Caroline Muir, renowned sexual & spiritual healer,
best selling author of *Tantra Goddess: Memoir of Sexual Awakening*

"Natalie Geld is doing more than anyone else I know to "lift the skirts of consciousness" and demand that we attend to the embodied, down and dirty aspects of transcendent reality. From singularity to sexuality, Nat walks the edge with brilliance and fearlessness."

–Cassandra Vieten, PhD, Clinical Psychologist,
CEO & President of the Institute of Noetic Sciences,
author of *Mindful Motherhood*

"Natalie Geld, a courageous and brilliant educator, has written a wonderful book that will transform your life. Her passion and commitment to healing sexuality and finding fulfillment in life is a shining beacon to all of us."

–Zoran Josipovic, PhD, neuroscientist,
Psychology Department, Center for Neural Science, New York University,
founder of the Nonduality Institute

"We live in a society that imprisons us on incompatibility row: the physical is incompatible with the intellectual, the sexual is incompatible with the spiritual, the pleasurable is incompatible with the moral. Natalie's work helps us break through the concertina-topped fences that maintain the segregation -- separation -- of our minds and bodies."

–David Lawrence, journalist, scientist,
author of *Upheaval from the Abyss:*
Ocean Floor Mapping and the Earth Science Revolution
and *Huntington's Disease*.

"Finally a book whose time has come! Author and educator Natalie Geld dares to go where few have gone before... out of the ashes of stifling cultural and religious taboos regarding sex straight into the reader's own "stories" about their body and mind. Liberating, humorous and self-empowering, her book is just what the doctor ordered."

–Christine Blosdale
award-winning critic & writer,
senior producer, KPFK 90.7FM

"Protective armor has its place, particularly in hazardous environments. However its greatest value is recognizing when you no longer need it. Natalie Geld navigates this paradigm like a laser."

–Jeremiah Sullivan, marine biologist, shark authority,
founder of Neptunic & SharkArmor Tech

"We aren't walking around as brain stems. Our mind is not separate from our body, our sexuality not separate from our heart and spirit – it's all combined, unified. Sensuality plays a huge part in who I am as a person, who I am as a writer, definitely who I am as a poet. For women, it's important to cultivate sensuality – because it's a vital part of what creates us, of who we are. As adults, we do so much to make what we say more pleasing to another's ear as opposed to laying it bare. Read this book and absorb its wisdom – read your own stories then lay it bare...get the darkness out. Once it's out, it's not lurking in your subconscious anymore. It's not shrouding your conscious mind, it's just out; it's gone."

–Azure Antoinette,
poet/spoken word artist, youth literacy advocate,
commissioned by Maria Shriver,
Beats By Dr. Dre & Girl Scouts of America

"There has never, in the history of science and academia, been a catalyst for outreach and education progress like Natalie Geld. Her magnificent work of raising the bar for communication and collaboration amongst the mind brain sciences, the media and creative arts, and the scientific study of consciousness is unprecedented."

–Bernard J. Baars, PhD
cognitive neuroscientist,
CEO/co-founder of
Society for Mind Brain Sciences,
affiliate fellow of The Neurosciences Institute,
author of *In the Theater of Consciousness: Workspace of the Mind*
and *Introduction to Cognitive Neuroscience*.

ACKNOWLEDGEMENTS

To convey my excitement for the beauty and importance of science with simplicity and conversational ease, missing are academic references, footnotes and such. I've logged a boatload and am happy to share if you're interested. I appreciate and respect the rigor and integrity involved in developing science, philosophy, and medicine. Any misrepresentations are unintended. I hold all my contributors in the highest regard.

Thank you with all my heart -- Caroline Muir, Rick Hanson, Cassandra Vieten, Zoran Josipovic, Gil Hedley, Jay Giedd, Kevin Krycka, Dean Radin, Anne-Marie Duchêne, Maurizio and Zaya Benazzo for your unique genius, warm welcome and abiding support.

Dr. David Kipper, for sparking this dialogue with me to set these wheels in motion. Stephanie Moses for your lifetime of love and invaluable insights. Mike Kerhin for your true friendship and production quality. Anastasia Shepherd for our early video work and generosity. Elizabeth Zack for your editing prowess helping me to become a better writer.

Bernie Baars, thank you… always thank you.

I'm incredibly grateful for -- My editor, David Jon Peckinpaugh for your wizardry, calling me out, and creatively brainstorming these books to life!

Tonietta Walters, Charles Ng'ang'a, DJP, Vanessa McNeill, and Christyna Jauffret for choosing to evolve beautifully together.

Dan Fesman, Deanna Robertson, Felecia Faulkner, Frederic Thomson, Becky Riley Fisher, Claire Waismann, Dean Holtermann, Christine Blosdale, Steve Miller, Robert Dahey, David S., JJ, John Lau, Robbie Simpson, Mike Farr, Diane Lane and my loving buds for having my back and front all these years.

The extraordinary Steven Lee Burright for shifting communication every day, cause right!?!

Alfie Rustom for supporting this vision wholeheartedly. Alison Anthoine for being my super powered legal eagle and dear friend. Sallie Thurman for your invaluable feedback.

Barry Wells for being the most amazing creative design god!

For my Daddy, even though he'll never read this, lol.

My brilliant kids, of course, whom I adore... Robbie, Ellarie, Christina, and Milli Melle for choosing me as their momma. I love you infinity bunches forever.

And for you – for always inspiring me, thank you!

RECOMMENDED RESOURCES

RickHanson.net

RickHanson.net/books

NondualityInstitute.org

Noetic.org

Noetic.org/education/self-study/
mindful-motherhood-course

Focusing.org

DeanRadin.com

DrLouann.ning.com

GilHedley.com

ArtOfAlignment.com

ScienceAndNonduality.com

Divine-Feminine.com

WhyAreWeWhispering.com

SpeakYourBodysLanguage.com

CONNECT WITH NATALIE GELD

Sign up for Nat's newsletter at

NatalieGeld.com

Connect with her on

Facebook.com/NatalieGeld

Twitter.com/NatalieGeld

Instagram.com/NatalieGeld

NatalieGeld.tumblr.com

GoodReads.com/Natalie_Geld

Pinterest.com/NatalieGeld

Google+NatalieGeldAuthor

ABOUT NATALIE GELD

Natalie Geld is Co-founder and Creative Director of the Society for Mind Brain Sciences. She has been actualizing vivid human potential for 25+ years. She is Founder of WhyAreWeWhispering.com, read in 202 countries and every state in America. Natalie is also producer/director of the PBS series "The Feeling Brain," and co-author of a book with cognitive neuroscientist Dr. Bernard Baars about one of the most interactive structures in the universe – our brain. Their book is also the story of the explosion of discoveries in consciousness science these past 15 years, published by Oxford University Press.

If you've seen Nat on TV or in magazines, it's because she straddles a diverse production career behind the camera, as well as an established career as an actress, model, and voice over artist with agencies like Ford and Innovative Artists NY. When Nat's supposed to be schmoozing, she rogues about instead asking, *"Why do we feel this way?"* and *"How can we evolve?"*

Author, Natalie Geld

www.NatalieGeld.com

Also available as an eBook and on audio from Nautilus Press.

CPSIA information can be obtained at www.ICGtesting.com
Printed in the USA
BVOW08s1939110916

461822BV00001B/1/P